GETT

With The

ANSWERING MACHINE

and

Voice Mail

JOHN CARFI & CLIFF CARLE

Illustrated by
Desmond Mullan

CCC PUBLICATIONS

Published by

CCC Publications
1111 Rancho Conejo Blvd.
Suites 411 & 412
Newbury Park CA 91320

Manufactured in the United States of America

Cover © 1995 CCC Publications

Interior illustrations © 1995 CCC Publications

Cover & interior art by Desmond Mullan

Interior production by Oasis Graphics

Cover production by Desmond Mullan

ISBN: 0-918259-99-1

If your local U.S. bookstore is out of stock, copies of this book may be obtained by mailing check or money order for $5.95 per book (plus $2.50 to cover postage and handling) to:
CCC Publications; 1111 Rancho Conejo Blvd.;
Suites 411 & 412; Newbury Park CA 91320

Pre-publication Edition - 10/95

First Printing - 3/96

CONTENTS

INTRODUCTION

Let's face facts — most answering machine messages suck. It's bad enough the person you're trying to reach is *out to lunch,* but to add insult, you're stuck with a #@??%*!! machine! Furthermore, your chances of hearing a clever or funny outgoing message are about as good as hitting the lottery—one in a million. The 999,999 other times you get a dopey, cutesy, or incredibly long-pointless-and-boring outgoing message. So as the old saying goes . . . "don't get mad—get even with the answering machine!"

AND THEN THERE'S #@??%*!!, #@??%*!! VOICE MAIL . . .

Few things in this world are more annoying than Voice Mail. It's torture enough that you usually have to sit through five minutes of infuriating and seemingly endless menu options: "If you want Accounting, press *one* . . . If you want Personnel, press two . . . If you want to speak to a real, live person, press your rear-end into your car seat and drive down here . . ." Then, after you've finally pushed the interminable series of numbers to get to the person you need to talk to, THEY'RE OUT! Or, in many cases, they're too lazy to pick up the phone so they leave their Voice Mail switch on all day! Well, here's the good part: those three to five minutes of sitting through menu options will give you plenty of time to look up the appropriate Getting Even message to leave, and then . . .

IT'S PAYBACK TIME!

HOW TO USE THIS BOOK

A. If you know when the person you are calling is usually out, call with a specific message ready. If you do reach a real person, save it for next time.

B. If you reach a machine unexpectedly, leave a normal message (name and number), then call right back with a gag message from this book.

NOTE: For easier readability, we have used the generic names "JANE" and "JOHN." (or "Ms. JANE" and "Mr. JOHN") Just substitute the machine owner's first or last name and/or your name where appropriate. Also, we have used a generic telephone number (123-4567) to leave as part of your gag message. Substitute your own number if the person you are calling has a sense of humor (or if your health insurance is paid up).

AUTHOR DISCLAIMER: "Getting Even With The Answering Machine (and Voice Mail) was written as a TOTAL GOOF. The main purpose is for EVERYONE to have fun. So, if you reach one of those "way serious" machine owners, we suggest you add the tag "just kidding" or call back and let them know you were only joking.

ALL "700" NUMBERS USED IN THIS BOOK ARE *FICTITIOUS!*

They were created for comic effect only. If any are real, working numbers it is purely coincidental and is not the authors' intention that they should be called.

When saying the "700" numbers, it's best to *spell them out*. That way, the machine owner will see he's been duped as he writes it down (e.g. 700 I-M-A — D-O-R-K)

Got it? Great. Now, GO GET 'EM!

CHAPTER 1:
FAMOUS MESSAGES

This is GENERAL NORMAN SCHWARTZKOPF. Take my word for it . . . your message bombed!

∗ CLICK ∗

This is DR. KERVORKIAN. My god! That's the worst message I ever heard—I almost *died!* Do you mind if I use it as a back-up in case my suicide machine malfunctions?

∗ CLICK ∗

Hello. This is THE PSYCHIC NETWORK with your daily horoscope: "You are gullible, naive, and often let people take advantage of you!" That concludes your reading. Now send us five hundred bucks.

∗ CLICK ∗

This is PRESIDENT CLINTON. Your message smells so bad that for the second time in my life, I made a point *not to inhale!*

* CLICK *

(ANGRY)
This is TANYA HARDING. After that message I'm starting to think the wrong person got whacked in the shins.

* CLICK *

(SLOW)
This is TOM HANKS. Answering machine messages are like a box of chocolates . . . you never know what you're gonna get . . . in this case . . . sick!

* CLICK *

This is JEFF FOXWORTHY. You might be a redneck if . . .No—for sure! By that message, I can tell you're *definitely* a redneck!

* CLICK *

This is DAVID LETTERMAN with the Top Ten reasons why you shouldn't be allowed to have an answering machine, but I'm gonna skip the first nine and give you the Number One Reason . . . (YELLING)
You're a bonehead and have no talent!

* CLICK *

(SHORT SCREAM)
This is STEPHEN KING. That message is so bad, it's scary!

* CLICK *

(DISAPPOINTED)
This is RICH LITTLE. Darn! I was going to do an impression of a jerk, but you beat me to it!

* CLICK *

Hello. This is the SURGEON GENERAL calling. I have determined that your message is hazardous to everyone's health!

* CLICK *

(ENGLISH ACCENT)
This is HUGH GRANT. That message sucks!

* CLICK *

This is the TERMINATOR. Your message blows! Don't worry, I won't be back!

* CLICK *

(HAPPY VOICE)
This is DR. JECKLE—I love your message!
(ANGRY VOICE)
This is MR. HYDE—I hate your message!

* CLICK *

[OPTION: MAKE TWO CALLS]

Uh, Hi. Uh, this is THE AMAZING KRESKIN. Uh, I'm really sorry—I must have dialed the wrong number.

* CLICK *

This is DIRTY HARRY. I don't know if that was one beep or two. But beings this is a 44-magnum, the most powerful handgun in the world, you have to ask yourself one question, "Do I want to change my message?" Well, do ya, Punk?

* CLICK *

Hi there! This is RICHARD SIMMONS. Please be a sweetheart and give me a call at 1 and 2 and 3 and 4 . . . and 5 and 6 and 7 . . . Come on now! let's sweat that ugly cellulite off those index fingers! That's 1 and 2 and 3 and . . .

* CLICK *

(SARCASTIC)
Hello. This is DR. JARVIC. What an inspiring message! I'll keep it next to my hearts!

* CLICK *

(SARCASTIC)
Hello. This is MARIE ANTOINETTE. What a funny message! I laughed my head off!

* CLICK *

This is COURTNEY LOVE. Dude, that message belongs in a deep HOLE!

* CLICK *

This is JOHN DALY (OR ANY OTHER FAMOUS GOLFER). Between this and your last message, I'd like to shoot a hole in one!

* CLICK *

This is NEWT GINGRICH. Just as soon as I finish out my "Contract With America," I'm gonna put a *contract* out on your machine.

* CLICK *

Hi. This is JENNY JONES. This week on my show we are featuring "Midgets Who Sleep With Megalomaniacs," "People With Hare Lips Who Have No Hair," and "Losers Who Put Stupid, Boring Messages On Their Machines!" So, Ya wanna be on my show?

* CLICK *

This is BILLY GRAHAM. After hearing that message, I'm going to pray for your sanity!

* CLICK *

This is JIM CARREY. If I had that message on *my* machine, I'd be afraid to go out in public without wearing a MASK. Man, what have you been SMOKIN'?

* CLICK *

This is SENATOR PACKWOOD — You must have bent over backwards to come up with that—but even I wouldn't *touch* that message if I were you!

* CLICK *

This is MICHAEL JORDAN. That message is *foul!* Do you always *dribble* when you talk? Your machine should be *slam dunked!* That's it, I need to go out for some *air!*

* CLICK *

This is SANDRA BULLOCK. Your message and that machine oughta be run over by a SPEEDing bus.

* CLICK *

This is JOHNNY CARSON. That message is so bad . . .

(PERSON IN BACKGROUND)
How bad is it?

(ala JOHNNY)
It's *so* bad, it would have to *improve* to "suck!"

* CLICK *

(THREE PEOPLE TALKING)
This is ROSEANNE, CHRIS FARLEY and DELTA BURKE. If you don't change that message we're gonna come over to your house and *sit* on your machine!

* CLICK *

TO A PERSON WHO *TRIES* TO PUT A CUTE OR CLEVER OUTGOING MESSAGE ON THEIR MACHINE:

(TWO PEOPLE — SARCASTIC VOICES)
This is SISKEL and EBERT. After reviewing your message, guess which way our thumbs are pointing?

* CLICK *

AFTER CORNY OR CUTESY MESSAGE:

Oooooo! Icky-yuck! This is ALICIA SILVERSTONE—and they thought I was CLUELESS!
Bite me, Dude!

* CLICK *

TO A FEMALE:

Hello. This is SANTA CLAUS. I called to let you know I'm doing things differently this year. You better be naughty if you want me to be nice!

* CLICK *

CHRISTMAS CALL:

Ho! Ho! Ho! This is SANTA!
You better not pout,
You better not cry,
You better watch out,
I'm telling you why:
Your message sucks reindeer doo-doo!

⁎ CLICK ⁎

[IMITATE *ANY* CELEBRITY]

Hi JOHN. This is (CELEBRITY'S NAME). I was
given your name by a mutual friend. Sorry you're
not there. I was going to invite you to my wild
party, tonight. All the *big names* are going to be
there. I'd leave my number, but, of course, it's
unlisted.

⁎ CLICK ⁎

CHAPTER 2

RATED "PG"

Hi! I'm so stoked! This is, like, my first time ever leaving a message on a machine (voice mail)! I hope I'm good and live up to your expectations. Well, here I go:
My number is 123 . . .
(EMBARRASSED)
Whoops, sorry! Premature message!

* CLICK *

Hello? Uh, listen, this is the apartment manager. Uh, due to numerous complaints from your neighbors, I'm going to have to ask you to put shock absorbers on your bed!

* CLICK *

Hey JOHN, I don't mean to freak you, man, but there's a rumor goin' round that your sex life is pretty lousy these days—guess you'll just have to take matters into your own hands!

* CLICK *

Hi. I need your advice. I'm a 65-year-old flasher. I've been exposing myself in public for the past 30 years. What do you think—should I retire? Or stick it out another year?

* CLICK *

Hello. This is your Pharmacy calling. I want to confirm an order we received earlier today for a case of "Preparation H." Now, it says here that you want it delivered to your back door???

* CLICK *

Hello. This is the Sperm Bank. Due to the infrequency of your deposits, we're going to have to close your account.

* CLICK *

(MECHANICAL VOICE)
Hello? JOHN'S answering machine? This is JANE'S answering machine. I hear you're a real electronic stud and generate a lot of voltage. Tell ya what—I'll show you my tape if you show me yours!

* CLICK *

Hi, JANE. This is JOHN. The good news is, last night you were fantastic! The bad news is, I'm calling from the clinic.

* CLICK *

(SEXY VOICE)
Hi there, you sexy hunk . . . Mmmmmmmm!
I'll bet this sounds like a real sensual, *orgasmic*
message . . . Ahhhhhhh! But I'm only faking it.

* CLICK *

Is this JOHN? This is your dentist calling. I wanted
to leave a reminder that Wednesday at 5:00 I'm
supposed to drill your wife.

* CLICK *

Hi there. I got your number off a restroom wall
and I'm definitely down for the "kinky sex" you
promised—but I'm a little freaked about the gig
with the "live porcupine." Grodo!

* CLICK *

[OPTION: HAVE A COUPLE FRIENDS CALL
WITH A SIMILAR MESSAGE]

(MECHANICAL VOICE)
Hello? JOHN'S answering machine? This is
JANE'S answering machine. I get so excited when
we make contact. My transistors turn red. My
capacitors enlarge. And my diodes get so hot. Oh
god! I need you to come over and touch my
knobs. Please hurry—and don't forget to erase this
message so your owner doesn't find out about us.

∗ CLICK ∗

(A COUPLE DEEP BREATHS)
Hhhhhhhhhhh . . .   Hhhhhhhhhh . . .
Hi. I'm returning your heavy breathing. And by
the way, next time, try a breath mint!

∗ CLICK ∗

Hello. This is JANE, the girl you met at that
wild, surprise party eight months ago. Well, in
about a month, I'm gonna have another
surprise for you . . .

∗ CLICK ∗

CHAPTER 3:
TRICKS 'R' US

*(Gags & Practical Jokes To Really Mess
With Their Minds)*

[AUTHORS' NOTE: PLEASE HAVE A HEART—
CALL BACK AND SAY "JUST KIDDING!"]

(RADIO ANNOUNCER VOICE)
Hello and congratulations! This is "Dialing For
Dollars!" $1000 is yours and all you have to do is
call us back at 123-4 . . . (MUMBLE). That's
123-4 . . . (MUMBLE). Call now!

＊ CLICK ＊

Hello. This is the No-Tell Motel. Congratulations,
"Mr. John Smith!" You've won our "Customer Of
The Year" Award!

＊ CLICK ＊

Here JANE. Machine your, funny sounds. Better it, take fixed, to be. Talk to, later you.

* CLICK *

(FOREIGN ACCENT)
Hello. This is Jose's Gardening Service. We grow the best roses 'cuz we use the freshest manure. I am leaving a sample in your mailbox.

* CLICK *

(RAPID POUNDING NOISE IN B.G.)
Yo, JOHN. I' calling from a phone booth and I . . . uh, hang on . . .
(CALLING OUT)
Chill out, man!
(POUNDING NOISE GETS LOUDER)
Me again, sorry JOHN! Uh, listen, there's a dude outside with a big red "S" on his chest—he wants in here real bad! Check ya later!

* CLICK *

IDENTIFY YOURSELF AND START TELLING
YOUR FAVORITE JOKE—BUT CUT YOURSELF
OFF (RAPIDLY CLICK THE DISCONNECT
BUTTON 3-4 TIMES) AND HANG UP JUST
BEFORE THE PUNCHLINE.

* CLICK *

(VERY FAST)
Hi, JANE. I'm rushing out the door, but I wanted
to let you know that a bunch of us are going out to
an awesome, new hangout tonight. It's primo:
terrific food, mondo band, dancing and plenty of
hot-looking hunks. So if you're interested, meet us
there.

* CLICK *

Hey JOHN, I know you're hosed about that money I owe you, so I'm calling to let you know that I'll be paying it back in *three* days . . . January 7th, July 8th, and December 9th.

* CLICK *

[OPTION: CALL A BUSINESS CREDITOR—OR COLLECTION COMPANY—WHO HAS VOICE MAIL—AFTER HOURS.]

(NORMAL VOICE)
Hello? Mr. JOHN? I am a lawyer with the FCC. I'm calling to inform you that the "beep" tone on your answering machine is in violation of code 3527.9—your particular frequency has been found to cause immediate sterilization in certain men . . .
(HIGH VOICE)
If you do not take your machine in and have the "beep" frequency changed, we will be forced to prosecute. Thank you for your cooperation.

* CLICK *

(INSECURE VOICE)
Uh, Hello? This is the, uh, Paranoid Joke Service.
Uh, Today I have a real funny joke, but if I told
you, you'd only laugh at me!

* CLICK *

HELLO-hello, JOHN-john. THERE-there SEEMS-
seems TO-to BE-be A-a STRANGE-strange ECHO-
echo IN-in YOUR-your MACHINE-machine!

* CLICK *

Hi. I called because . . .
(PAUSE)
Uh, I forgot???
Anyway, call me back later—if I remember.

* CLICK *

OFFICE GAG:

DAY I:

HAVE 3 OR 4 CO-WORKERS CALL YOUR
FRIEND'S MACHINE AND AD LIB, FOR
EXAMPLE:

- Hello? Hello? Hello?

- Hello? Whose machine did I reach? I can barely
  hear you!

- Say what? Louder, Dweeb!

- What're ya whisperin' for? Speak up!

  ETC.

DAY II:

MOST LIKELY, BY THE NEXT DAY, THE MACHINE OWNER WILL HAVE RE-RECORDED HIS MESSAGE, A LITTLE LOUDER THIS TIME. HAVE 3 OR 4 CO-WORKERS CALL AND AD LIB, FOR EXAMPLE:

• Hey man, why is your message so loud?

• Chill out, Dude! You almost popped my eardrum!

• Yo, Dufus! Ya got yer volume maxed out!

• What're you, a noise freak? *Can* the yellin', Retard!

ETC.

CALL #1:

Hello. This is the laundry—we lost your shirt!

∗ CLICK ∗

CALL #2:

Hello. This is your broker—you lost your shirt!

∗ CLICK ∗

CALL #3:

Help! This is your shirt—I'm lost!

∗ CLICK ∗

Hi. You probably don't realize it, but I'm performing an incredible magic trick on your machine right now. I'm actually leaving two messages at once . . . my first—and my last!

✻ CLICK ✻

(PLACE HANDKERCHIEF OVER YOUR RECEIVER, OR ALTER YOUR VOICE)
Hi. Can you guess who this is?
Nope! guess again . . .
Nope! guess again . . .
Nope! guess again . . .
Aw, that was close! Guess again . . .
Nope! guess again . . .
(CONTINUE FOR AS LONG AS YOU WANT—EVEN CALL BACK AND CONTINUE IF THE MACHINE HAS A LIMITED TAPE AND CUTS YOU OFF.)

✻ CLICK ✻

TO A PET OWNER:

Hello? Mr. JOHN? This is the Biological Testing Facility. Listen, we're terribly sorry to report that one of our trucks broke down in front of your house this morning and released a couple million laboratory fleas onto your lawn. But don't worry, as soon as our Lab Assistants get back from their vacations, we'll send them right over.

* CLICK *

[NOTE: CALL WHEN YOU KNOW THE WIFE WILL PICK UP THE MESSAGES BEFORE THE HUSBAND DOES — CALL BACK IMMEDIATELY AND SAY "JUST KIDDING!"]

Mr. JOHN? This is the [LOCAL NAME] Dry Cleaners with some bad news. We're very sorry, but when we cleaned your jacket, we didn't notice the black lace see-through panties that were stuffed into the pocket and they got ruined. We'll be glad to work something out on your bill, okay?

* CLICK *

(PAUSE BRIEFLY ON EACH DASH)
Hello? JOHN? This is —ANE. I'm not —ure, but I
think there's —omething wrong with your
—achine. Anyway, give me a —all at one, two,
—ree, four, —ive, —ix, —even (OR YOUR
NUMBER LEAVING OUT THE FIRST
CONSONANTS). Thank —ou!

* CLICK *

Hey, Bro. I was calling to tell you about this
wicked book that's out, "How To Torture Your
Friends And Relatives," but from your message, I
can see you already got it.

* CLICK *

Hi. I'm a professional mind reader. I charge $100
for geniuses, $75 for above average and $50 for
average. But based on that message, for you,
consider it a freebie. Call 700-HALF-WIT.

* CLICK *

[HAVE A FRIEND MAKE BARKING NOISES IN
THE B.G. AS YOU START TALKING]

Hi JOHN, it's JANE. This is just a social call, so
when you get a chance . . .
(DOG BARKING)
Excuse me a second . . .
(TALKING AWAY FROM THE RECEIVER)
FIDO! (OR YOUR DOG'S NAME) How many
times have I told you—that's my business phone!
Use the phone in your dog house!
(BACK INTO THE RECEIVER)
Sorry, JOHN. Anyway, call me, okay?

＊ CLICK ＊

This is the Hot-Tip Money Line! Dial us up, have
your credit card ready and you'll realize thousands
of dollars in return with our great money-making
tips. To whet your appetite, here's a freebie:
Go into a furniture store. Put a chair or something
on layaway with a small deposit. Here's the in-
genious part: go back 100 years later and pay the
balance! Now you've got an antique—worth five
times the original price! Call now! 700-IMA-FOOL.

＊ CLICK ＊

Hi. Sorry to bother you, but I have amnesia and
I just called to say . . .
(PAUSE)

* CLICK *

(CALL BACK—SAME TONE OF VOICE)
Hi. Sorry to bother you, but I have amnesia and
I just called to say . . .
(PAUSE)

* CLICK *

(CALL BACK—SAME TONE OF VOICE)

Hi. Sorry to bother you, but I have amnesia and
I just called to say . . .
(PAUSE)

* CLICK *

(Continued on next page)

34

(CALL BACK—SAME TONE OF VOICE)
Hi. Sorry to bother you, but I have amnesia and
I just called to say . . .
(PAUSE)

* CLICK *

(CALL BACK—SAME TONE OF VOICE)
Hi. Sorry to bother you, but I have amnesia and
I just called to say . . .
(PAUSE)

* CLICK *

(CALL BACK—SAME TONE OF VOICE)
Hi. Sorry to bother you, but I have amnesia and
I just called to say . . .
(PAUSE)

* CLICK *

[NOTE: QUIT ONLY WHEN YOU FIGURE YOUR
FRIEND HAS HAD ALL HE/SHE CAN TAKE]

THE MAD JOKESTER

*MAXIMUM TORTURE: THE WORST OF
THE WORLD'S WORST JOKES!*

[DISGUISE YOUR VOICE—REALLY HAM IT UP!
AND CALL EVERY DAY FOR A WEEK—OR
UNTIL THEY PLEAD FOR PITY]

DAY #1:

Hey! It's the **Mad Jokester** on the loose!
Why wouldn't the lady let her doctor operate on
her husband?
She didn't want anyone to open her *male!*
Hey! What a *cut-up!*

* CLICK *

DAY #2:

Hey! It's the **Mad Jokester** again!
What do you get if you crossed the Atlantic with
the Titanic?
You'd get *halfway!*
Hey! That joke's *all wet!*

* CLICK *

DAY #3:

Hey! It's your lucky day! The **Mad Jokester's** back!
Why was the wagon train stuck in the middle of
the prairie?
It had *Injun* trouble!
Hey! I oughta be *scalped* for that one!

* CLICK *

DAY #4:

Hey! It's your ol' pal, The **Mad Jokester!**
What do you get when you cross a lighthouse with
a henhouse?
Beacon and eggs!
Hey! I'm *cookin'* now!

* CLICK *

DAY #5:

Hey! **Mad Jokester** checking in!
What do you get when you cross a praying mantis
with a termite?
A bug that says "grace" before eating your house!
Hey! That one *bit the dust!*

* CLICK *

DAY #6:

Hey! Heeeeeeeeeeeeeeeeer's Mad Jokester!
What do you get when you lean a corpse up
against a doorbell?
A *dead ringer!*
Hey! That was *cold!*
After that one, I think I need a *stiff* drink!

∗ CLICK ∗

DAY #7:

Hey! I think we had a lot of laughs, but the
Mad Jokester has other buddies to cheer up.
But before I go . . .
Did you hear that Wurlitzer is merging with Xerox?
They're gonna make *reproductive organs!*
Whoa!. that's playing *dirty!*

Hey, you know you're gonna miss me when
I'm gone!

∗ CLICK ∗

Hi, JOHN? Did JANE tell you about me? I'm her friend, the *artist*. She said you were way cool, but after that message, I'm giving you the *brush*-off!

* CLICK *

This is City Hall calling. We've been notified that you have a birthday party coming up and we just wanted to inform you that, by law, when they light all those candles on your cake, a Fire Marshal must be present.

* CLICK *

[NOTE: MAKE ONLY ONE CALL]

(SOMEWHAT FRAZZLED)
Uh . . . I called to tell you that the plans for tonight have changed, but, on second thought, let's stick with the original plan I left in my *earlier* message. Umm, I'm not home, but you can still reach me at that number I left *earlier*.

* CLICK *

[NOTE: CALL WHEN YOU KNOW THE WIFE WILL PICK UP THE MESSAGES BEFORE THE HUSBAND DOES — CALL BACK IMMEDIATELY AND SAY "JUST KIDDING!"]

Hi JOHN, this is the new girl at the office. I've reconsidered and the answer is "yes!" And by the way, thanks for the promotion.

* CLICK *

Hey JOHN, want to hear the world's funniest Pollock joke? Just dial 976-POLZ.

* CLICK *

[NOTE: TRY IT YOURSELF, FIRST]

CHAPTER 4:

Expressly VOICE MAIL

This is the [NAME BRAND] Aspirin Company. Congratulations Mr. JOHN, you're our "Man Of The Year!" We want to thank you for keeping us in business by being a royal "pain-in-the-A!"

* CLICK *

TO A "TYPE A" CO-WORKER:

[CALL *JOHN'S* VOICE MAIL — PRETEND YOU THINK YOU REACHED SOMEONE ELSE'S]

(WHISPERING)
Hello, JANE? About the merger—it looks like it's going through. Unfortunately, a lot of heads are going to roll. So, whatever you do, don't breathe a word of this to anyone—especially JOHN!

* CLICK *

[NOTE: BE SURE TO CALL "JOHN" BACK AT THE END OF THE DAY AND LET HIM KNOW YOU WERE PUTTING HIM ON SO HE DOESN'T HAVE TO INCREASE HIS ULCER MEDICATION.]

[TO A "PRUDISH" OFFICE WORKER — CALL THE COMPANY RECEPTIONIST]

Yeah, would-ja leave a message on Ms. JANE'S Voice Mail fer me? This is Bizzaro's Sex Shop. We got us a truck load-a vibrators in here today and we're tryin' ta straighten somethin' out on her purchase order, here. Ax' her which model she ordered: the rip-cord or the kick-start? Oh yeah, our number's 800-SEX-MEUP. Tanks!

✳ CLICK ✳

This is the Personnel Department. As you may or may not know, we do yearly evaluations to determine what each employee is best suited for. In your case, it's *retirement*.

✳ CLICK ✳

Uh, yes, MS. JANE? This is the Unemployment Office. We've heard the rumors, too. Thought you might like to get a jump on things. Call 700-YER-GONE!

✳ CLICK ✳

TO AN UPWARDLY MOBILE EXECUTIVE:

Hello, Mr. JOHN. I'm sorry I didn't catch you in. To be frank, I'm a Headhunter, so I can't leave my name or number, I'll have to try again later. But you'll be pleased to know that based on your credentials, I have a job opening where you'll be able to boast that you have hundreds of people under you. The position is "Cemetery Caretaker."

* CLICK *

TO THE HEAD OF A MAJOR CORPORATION:

(FOR BEST EFFECT USE A HEAVY, FOREIGN ACCENT)
Uh, yes, Mr. JOHN, this is copier repairman. I am cleaning copier the other day an' someone left original document in machine. By mistake I put it in with my paperwork. Do you want me mail it you? Or bring it with next time I come? I don't know if important, but it say "resume" at top and I think it belong to one of your vice presidents. You can please call me at 123-45 . . .
(CUT YOURSELF OFF)

* CLICK *

TO SOMEONE WHO WEARS A HAIR PIECE OR HAS HAD HAIR IMPLANTS:

Hey JOHN, I just want you to know I stuck up for you today. A couple of babes in the secretarial pool saw you walk by and were arguing whether or not that was *your* hair. I said, "Of course it's JOHN's hair—he has the paid invoice to prove it!"

✳ CLICK ✳

TO A NEW EMPLOYEE:

This is the Payroll Department. As you may or may not know, we do yearly evaluations to determine what each employee is worth . . . You should thank God for the Minimum Wage Law.

✳ CLICK ✳

Hello Mr. JOHN? I'm an editor with [BIG NAME PUBLISHING HOUSE]. I was referred to you by your company's Accounting Department. I'm prepared to offer you a hefty advance to write a book of fiction based on your expense account. Call 700 FAT-LIAR.

✳ CLICK ✳

BLUE SKY Products

[CALL ON CONSECUTIVE BUSINESS DAYS]

Hello. I'm a telemarketer with Blue Sky products — B.S. for short. Today's B.S. specials are:
- "Shake 'N' Bake" for sushi
- For really fat people, a bathroom scale made of Hershey's chocolate
- And "mono headphones" — for people with a one-track mind.

For more information, call 700-IAM-DUMB.

* CLICK *

Hello. It's Blue Sky products, again. Are you overweight and proud of it? Then you'll love these new B.S. non-health products:
- Food-flavor lip gloss
- Blueberry pancake make-up
- and our special of the month, "Oil of O'Large."

For details call 700-IMA-FATY.

* CLICK *

Hello. Blue Sky products. Is sex with your spouse like a ride on a roller coaster? It lasts two minutes and you get sick afterwards! Well, try the latest B.S. innovation: "TY-ONE-ON," the pain tablets that instantly *give you a headache!* That's "TY-ONE-ON," for the perfect excuse not to have sex! To order call 700-SEX-SUXS.

* CLICK *

Hello. It's your new friend, Blue Sky, again. Do you hate exercise? Need an excuse *not* to go jogging? Our latest product is a transistor radio with 500 pound headphones. It's the Sony "I-Can't-Walkman." Get out your credit card and call 700-JOG-NOMO.

* CLICK *

Hello. This is the Organ Transplant Bank . . . your brain is ready.

* CLICK *

Mr. JOHN? This is your Financial Planner. *Plan* on working weekends at the 7-Eleven. Me, your money, and your girl are off to Brazil.

* CLICK *

(TV ANNOUNCER VOICE)
Hello Mr. JOHN. We were calling you today for your chance to win big bucks playing "Telephone Jeopardy." Every day we call someone up and challenge them to a "battle of wits" — but after hearing your message, I can see you're unarmed.

* CLICK *

This is the bank calling. Your wife's check for a one-way ticket to the Hedonism Retreat in Jamaica has bounced.

∗ CLICK ∗

This is the bank calling. Your husband's check to 900 HOT-BABE has bounced.

∗ CLICK ∗

This is your bank calling. Your account reminds us of a high-fashion model's chest: flat busted!

∗ CLICK ∗

(PROFESSIONAL RADIO VOICE—FOR BEST EFFECT, START TALKING JUST AS THEIR OUTGOING MESSAGE IS FINISHING)
. . . 4567! Dial that number now and find out about the brand-new Mercedes you just won!

* CLICK *

Hello? Mr. JOHN? This is Webster's. The word is you're pretty smart. So maybe you can tell us . . . if a word in our dictionary was mispelled, how would anyone know?

* CLICK *

(PROFESSIONAL SOUNDING VOICE)
This is Computerized Career Planners, a non-profit agency. Using highly sophisticated instruments, we are able to analyze your voice print and determine what career you are best suited for. Based on your message, our printout shows that you should be . . . a bone specialist — you definitely have the *head* for it. For more information call 700 NO-BRAIN.

* CLICK *

Hello? Ms. JANE? This is the [LOCAL NAME] Beauty Parlor. One of your co-workers sent us your picture for a makeover evaluation. Uh, we'll have to get back to you—it's gonna take a few more days to do the complete estimate.

* CLICK *

TO A PERSON WHO GENERALLY TALKS AND MOVES SLOWLY:

Hello. This is the Post Office. If you're interested, we have a job opening. One of our employees came out of his coma, so we had to let him go. Call 700-WE-GO-SLO.

* CLICK *

Hello. This is the IRS. We know how much you make. We were wondering what it must be like to have all that money—and come April 15th, we're going to find out!

* CLICK *

MAXIMUM VOICE MAIL PAYBACK:

(FAST)

Is this (EMPHASIZE THEIR *FIRST & LAST NAME*)? I'm a private detective. Your name came up in a will and you stand to inherit a sizable fortune if you are *the* (FIRST & LAST NAME). Call me immediately at 123-4567. If I'm not there, try 061-0728. If that number is busy, try 102-1863 or 026-5180. Now if you get my service, hang up and dial 172-8750 or 057-2841—no, better yet, try 191-3572. If I'm not there, try my "700" number . . . Uh, let's see, oh yeah, it's 700-URA-JERK.

∗ CLICK ∗

CHAPTER 5

SPECIAL DELIVERIES

(For "Special" People or Typical Outgoing Messages)

AFTER "LEAVE YOUR NAME, NUMBER AND A MESSAGE":

Uh, I'm not sure who I am—I forget my number—
and my message is a secret, so I can't tell you.

∗ CLICK ∗

AFTER "LEAVE YOUR NAME AND NUMBER":

My name? JANE.
My number? I'm a "10."

∗ CLICK ∗

TO A PERSON WHO LEAVES THE SAME MESSAGE ON THEIR MACHINE:

They say crime doesn't pay? Well Dufus, I'll give you ten bucks to change that message—it's killing me!

* CLICK *

TO A PERSON WHO LEAVES THE SAME MESSAGE ON THEIR MACHINE:

Hey, Dude, What's freakin' goin' on? Every time your message plays, dust and a musty odor comes outta my receiver . . .

* CLICK *

TO A PERSON WHO LEAVES THE SAME MESSAGE ON THEIR MACHINE:

Guinness calling. Okay! Okay! You did it. You've set the world's record for "longest running answering machine message." So change it already!

* CLICK *

AFTER A CORNY OUTGOING MESSAGE:

(ROSEANE ROSEANADANA VOICE)
That message was so funny I forgot to laugh.
I also forgot to leave my name and number.

* CLICK *

AFTER A CORNY OUTGOING MESSAGE:

Ugh! Ga-ross! You oughta provide noseplugs with
that message!

* CLICK *

AFTER A CORNY OR BLAND
OUTGOING MESSAGE:

Hi. You know that old saying, "If you don't have
anything nice to say, don't say anything!"
(PAUSE—AS IF THINKING)

* CLICK *

AFTER A CORNY OR BLAND OUTGOING MESSAGE:

I was so depressed today . . .
At first I was gonna jump off a building and land on a bed of poison-tipped spikes. Or, maybe drink nitro-glycerine and dive into a vat of scalding, hot oil while holding a 10,000 volt electric wire in my teeth. Nah—not horrible enough! Then it came to me . . . "I know! I'll call JOHN'S machine and listen to his message!"

* CLICK *

TO A PERSON WHO SCREENS THEIR INCOMING MESSAGES:

[NOTE: WORKS BEST WITH 3-4 PEOPLE HELPING]

(MAKE MISCELLANEOUS PARTY NOISES: LOUD MUSIC IN B.G., LAUGHTER, CLINKING GLASSES, ETC.)
Hey JOHN, I called to invite you to a killer party over here at my new friend's house, but I guess you're out. Deal with it, Dude . . .
[HANG UP EVEN IF HE PICKS UP THE RECEIVER]

* CLICK *

TO SOMEONE WHO DOESN'T RETURN MESSAGES:

Hey JANE, you never return my calls. You know, I've figured out something you'd be *perfe*ct at . . . being a perfect stranger!

* CLICK *

AFTER A LONG OUTGOING MESSAGE:

(WEAK VOICE)
Your message was so long and boring I have just enough strength to hang up...

* CLICK *

AFTER A LONG OUTGOING MESSAGE:

(SNORING SOUND)
Zzzzzzzzzz . . . Zzzzzzz . . . Zzzzzzzzzz . . .

* CLICK *

AFTER A LONG OUTGOING MESSAGE:

You know, I kinda liked your message in the beginning, but by the end, you talked me out of it.

* CLICK *

AFTER A VERY LONG OUTGOING MESSAGE:

(CLAP HANDS A FEW TIMES)
 . . . Oh! You're finally through. By the way, that wasn't applause. I was slapping my face to stay awake.

* CLICK *

AFTER A VERY, VERY LONG AND VERY, VERY BORING MESSAGE:

You know, it use to be I didn't believe in Hell — but I just went through it!

* CLICK *

AFTER A VERY, VERY LONG AND VERY, VERY BORING MESSAGE:

I don't know if you have any enemies, but I'll say this . . . that message is a great way to get even with them!

✳ CLICK ✳

AFTER AN OUTGOING MESSAGE THAT'S HARD TO UNDERSTAND:

Hey Wastoid, I notice you're having trouble with your *vowels*—have you tried a laxative?

✳ CLICK ✳

TO A PERSON WHO LEAVES THE SAME MESSAGE ON THEIR MACHINE:

Ms. JANE? This is the Coroner's office. We think it's time you put that message to rest!

✳ CLICK ✳

TO A PERSON WHO CHANGES THEIR OUTGOING MESSAGE OFTEN:

Yo, Dweeb, I notice your message is different every day. It's good to see you don't make the same mistake twice!

✳ CLICK ✳

AFTER CALLING EACH OTHER'S MACHINES WITHOUT CONNECTING "LIVE":

Hi. JANE. I'm returning the call you just left on my machine from when I left a message on your machine the other day because of the previous call you left on my machine a couple days ago about the original message I left on your machine last week— only now I forget why I called in the first place???

✳ CLICK ✳

LATE NIGHT CALL:

This is the Cemetery calling. We just wanted to remind you that you have to be back before sunrise.

✳ CLICK ✳

TO A GUY WHO CONSIDERS HIMSELF A REAL "HE-MAN":

Hello. This is Walt Disney Productions. Are you busy today? Tinkerbell called in sick and we need a fairy.

* CLICK *

THE DAY AFTER A PARTY:

(GROGGY)
Hey JOHN, what a trippin' party last night! I never drank so much in my entire life! Anyway, Dude, I need a reality check: did I have a good time?

* CLICK *

TO A GUY WITH EXCESSIVE BODY HAIR:

Hello. This is the Darwin Foundation. We're currently running tests to prove conclusively that man evolved from apes. Several of your friends volunteered your name as a research subject. Please call 700-HAIR-BOY.

* CLICK *

65

TO A PERSON WHO IS OVERLY FASHION-CONSCIOUS:

Hello. This is the Ringling Brother's Circus. One of our clowns quit and we're offering you the job—not because you're funny—we've seen the way you dress! Call 700-UBA-SLOB.

* CLICK *

TO A PERSON WITH SLOW/DULL/OR BORING VOICE:

Yo! Clueless! Some free advice: I think you should either put more fire into your messages or more of your messages into the fire!

* CLICK *

TO AN EX-HIPPIE OR ANY RELIGIOUS PERSON:

Hello. This is the Phone Poll Service. You've been specially selected to answer today's survey question: If God took acid, would he see people?

* CLICK *

TO A PERSON WHO GETS GROSSED-OUT EASILY:

Hi. You don't know me. The reason I called is because I'm trying to break my nose-picking habit. My doctor said, whenever I get the urge, stick my finger in the nearest hole seven times.

* CLICK *

TO A WOMAN WHO HAS BEEN PUTTING ON WEIGHT:

Ms. JANE? This is the Census Bureau, 123-4567, with a quick question: Have you been raiding the frig a little too often lately? — or should we be counting you as *two* people?

* CLICK *

MESSAGE FOR A BALD FRIEND:

(READ LIKE A TELEGRAM)
Hi. I'm at the barber shop.
Having a great time.
Wish you had hair.
Yours, JOHN.

* CLICK *

TO SOMEONE WHO HAS A LOT OF ANTIQUES OR PRICELESS HEIRLOOMS:

Hi. This is a burglar. While you were out I broke into your house, stole something of value, and replaced it with an *identical* cheap replica—but you'll never be able to prove it.

* CLICK *

[NOTE: IF YOUR FRIEND DOESN'T HAVE A SENSE OF HUMOR, CALL THEM SOON AFTER TO EASE THEIR MIND]

TO A FRIEND WHO HAS A COLD:

(NERVOUS)
Hello. Mr. JOHN? I'm a new mortician just out of school and I'm trying to start a new cemetery. Anyway, I heard you're not feeling well and I really want to get this thing off the ground, so to speak—I hope I'm not being too pushy, but if you're not feeling better soon, how about dropping in for a free measurement, okay? Call 700-6FT-UNDR.

* CLICK *

69

TO A CAT LOVER:

Hello? This is the City Pound. 123-4567. We're trying to find a home for a cute little kitty-cat. He'll be no problem, he's an indoor/outdoor cat— he's also machine washable!

* CLICK *

TO A DOG LOVER:

Hello. This is the City Pound. 123-4567. We're trying to find a home for a cute little puppy—it's a Mexican Chihauhau—his name is "Future Burrito." He's very easy to train. Whenever he misbehaves, just hold up a tortilla.

* CLICK *

TO A REAL FLAG-WAVING PATRIOT:

Mr. JOHN? This is the CIA. Every so often, we enlist the aid of brave and loyal citizens such as yourself. We have reliable information that one of your neighbors has removed the illegal tag off his mattress and has been opening the wrong end of milk cartons. If you know who the perpetrator is, do not try to apprehend him yourself. Call 700-URA-BOOB.

* CLICK *

THE WEEK AFTER HALLOWEEN:

Hey dude! Saw you at a club last night. Dude, didn't anyone tell you—Halloween was *last* week!

* CLICK *

TO AN EGOTISTICAL WOMAN:

Hello? Ms. JANE? We're the TV Commercial Modeling Agency. One of your friends sent us your picture and they were right! You're just what we're looking for! We hope you'll come by our office in person because we can use you right away. Call 123-4567 for directions, and when you get here, be sure to go in the door marked "Before."

* CLICK *

TO SOMEONE WHO IS UNLUCKY (or, ANY GAMBLER):

Mr. John? This is the Weather Bureau. As a matter of policy, we like to hire people who always guess wrong. A number of your friends referred us to you. If you're interested in a job, get back to us at: 700 BAD-CALL.

* CLICK *

Chapter 6:

FAKE WRONG NUMBERS

*(Fake Out Your Friends By Pretending You
Dialed A Wrong Number!)*

Hello? Suicide Prevention? This is the third time
I've called without a response from you. If you
don't get back to me this time, I'm gonna do it! It's
Donald at 123-45 . . .
(CUT YOURSELF OFF)

∗ CLICK ∗

[NOTE: CALL YOUR FRIEND RIGHT BACK AND
CONFESS SO THE NEXT MESSAGE THEY HEAR
THEY'LL KNOW THE PREVIOUS CALL WAS A
PUT-ON.]

(BROKEN ENGLISH)
Hello, please. Is this the Consulate? I am in your country only few days and no do understand your customs. Can you tell me please why last night I have problem at dance club? I only follow instruction! I am dancing real goodly and someone yell, "get down and boogie!" So I lay on floor and pick my nose—they throw me out of club! Why please?

* CLICK *

Hi. You don't know me, but this weekend I'm holding a meeting at my house for people who suffer from paranoia. However, since I don't know you that well, I better not tell you my name or where I live.

* CLICK *

(DISGUISE YOUR VOICE — TIRED & WEARY)
Hi. You don't know me, this is a total stranger. I am recovering from a forefinger injury and as part of my prescribed physical therapy, once a day for the next month, I have to dial every number in the phonebook. Well, talk to ya tomorrow.

* CLICK *

(TALK VERY SLOW)
Hello? . . . Dr. Smith? . . . It's . . . JANE . . .
You . . . know . . . I . . . guess . . . I . . . should . . . have
. . . called . . . first . . . then . . . took . . . the . . .
tranquilizers . . .

* CLICK *

(SIMULATE BEING "STONED")
Hey Dude, this is great, man! Like, I was just sitting
here doin' some weed and, like, I came up with the
world's funniest joke: "Two guys walk into a bar—
no wait, five Nuns—no, fifty penguins —wait—it
wasn't a bar, it was a sperm bank, and they were
eating—no—they were snorting cream cheese—no,
wait . . .
(PAUSE)
uh, I'll get back to ya!

* CLICK *

(DRUNKEN VOICE)
Hi. (HIC) Who'z Thiz? (HIC) Thish is JANE (HIC)
I think??? Lissen, If you have any idea as to my
whereabouts (HIC), pleash give me a call . . .

* CLICK *

Hello? Is this Dr. Sister's Sex Line? This is "Mr. X." again. Listen, remember that problem I had with the Olympic swimmer? Every time I wanted to go to bed, she wanted to go swimming? Well, last night I got her to compromise—we jumped in the pool and spawned!
Anyway, what should we do? Her coach is threatening to kick her off the team 'cuz now she'll only swim in one direction . . . upstream!

* CLICK *

This is the [CITY NAME] Medical Lab. I'm afraid we have some bad news. After rigorous testing we've finally determined what that growth on your neck is . . . It's your head.

* CLICK *

Hello? Is this the Cable Company? This is JOHN at 123-4567. I was hopin' you could help me. If you eat a TV Dinner, then throw up, is it considered "bad reception?"

* CLICK *

Hello? City Florist? This is Mort. 123-4567. Can ya help me out? A friend-a mine's in the hospital—just had a transplant operation—they gave 'em one-a those artificial hearts. So I was wonderin'—would it be appropriate to send him artificial flowers?

* CLICK *

Hello? Bob's Appliance Center? Listen, I ran into my ex-wife at a party the other night. She said I was looking a little pale and a great way to get a real quick tan is to stick my head in the microwave. Anyway, I was wonderin' if you could tell me what setting to put it on? My name's Bud and my number is 123-4567. Thanks!

* CLICK *

Hello? Travel Planners? Listen, I have to go to Detroit and I read some statistics that said, "the odds of getting shot and killed there are 1 in 5—whereas, the odds of getting shot *twice* there are 1 in 50,000." So I'm thinking, as soon as I get there, maybe I should wound myself in the leg? Wha-da-ya think? This's Duke at 123-4567.

* CLICK *

Hi. You don't know me, but I just dialed your number with my nose. And now I'm going to show you another trick: I'm going to leave this message blindfolded and with both hands tied behind my back:
(PAUSE)
"Help! I've been robbed!"

* CLICK *

Hello. I'm a new plastic surgeon in town and I've just opened my clinic. I offer major discounts, but I still can't seem to get any business and I haven't a clue why??? But if you're interested, I'm offering a *free* consultation. Just call 123-4567 and ask for Dr. Potatohead.

* CLICK *

TO A REAL "HE-MAN" TYPE GUY:

Hello? Uh listen, I dialed your number by mistake, but I'm a casting director and you've got just the voice we're looking for. Please call 123-4567 right away! We're trying to fill the part of a female impersonator.

* CLICK *

Chapter 7:

PHONEy SOLICITORS

Hello and congratulations from your City Council!
You've been specially selected to participate in our
City Beautification Project. Your job is to keep out
of sight!

* CLICK *

Hello. I'm a scientist with the Encephalogy
Institute. Based on your message, you're a perfect
test subject. We're looking for a brain that's never
been used. Call 700-AIR-HEAD.

* CLICK *

Hello. This is the Qwik-Kill Poison Company.
We're calling to see if we could use your picture on
our bottles.

* CLICK *

Hello. This is the National Charities Commission. Congratulations! You've been named the poster-boy for Hemorrhoids!

* CLICK *

Hello. This is "Perfect Couple" Dating Service at 123-4567. You'll be pleased to know that we put your statistics into our computer and have come up with your perfect match. For your first date we suggest you dress casual, be on time, and if you plan to go to the park or beach, be sure to bring a leash.

* CLICK *

This is the City Zoo. We know where you're hiding. Are you going to come in quietly—or do we have to bring a net?

* CLICK *

Hello. I'm with the Casualty Company. We're offering machine owners such as yourself a new coverage called "Answering Machine Insurance." It will protect your machine at times like this when your outgoing message is a total loss. Call 700-STOO-PID.

* CLICK *

(SERIOUS, AUTHORITATIVE VOICE)
Hello. This is the Department Of Defense, telephone number 123-4567. Congratulations! You have been selected at random by our computer to participate in a, uh, "low level" nuclear test experiment which will be held in a remote area of Nevada this coming weekend. Please be ready to be picked up Saturday morning at 0-600 hours. We assure you it's completely harmless, but just to be safe, please bring your darkest sunglasses and # 1,000,000 sunblock.

* CLICK *

(SERIOUS, AUTHORITATIVE VOICE)
Hello. This is NASA. We are taking applications for civilian passengers on the next Space Shuttle— its mission: to try and make contact with life forms from other planets. Anyway, a friend of yours sent in your name and a current photo. We're sorry, but on seeing your picture we've decided to "pass" on you—we don't want to give alien life the wrong impression.

* CLICK *

Hello. This is the Oriental Laundry. So sorry, we lost your clothes—but we have many pictures of them!

* CLICK *

Hello. We're a new restaurant in your neighborhood called *Cannibal's.* We're inviting you to come in and sample some of our tasty "Specialties," such as . . .
• Elbows Au' Gratin
• Hand-On-Rye
• All-Caucasian Patty
• Ladies' Fingers
• Our ever-popular "Leg-Of-Larry"
• And for dessert, "Eyes-Cream!"
Remember, that's *Cannibal's*—the restaurant that brings new meaning to "Hot Cross *Buns!*" For reservations call 700-CUT-MEUP.

* CLICK *

(PROFESSIONAL SOUNDING VOICE)
Hello? Mr. JOHN? This is the Phone Poll Service. I'm taking a survey. How do you feel about the Arms Race being over? I don't know about you, but I'm stoked 'cuz after about 25 yards, my arms would just peter out!

* CLICK *

Hello. This is your local Police Department. Don't panic, but as a precaution, we are calling to warn shoppers about the armless pickpocket. This guy puts his mouth where your money is!

* CLICK *

Hello. This is the Sharp Mind Memory School. You just won a free six-week memory course.
To confirm, please call us at 123-45 . . .
(PAUSE)
Uh, let's see . . .
Is it 67? or 76???

* CLICK *

Hello. This is the Punk Rock Beauty Salon. You have won a free facial, including massage—but give us a couple of days, our air hammer broke down.

* CLICK *

(UP TEMPO)

Hi, Sport! This is the Barnum & Bailey Circus. Say, Sparky, when you were young, did you ever fantasize about joining the circus when you grew up? Well Spanky, now's your chance—our geek quit! For information call 700-DUMB-ASS.

* CLICK *

Hello? Mr. JOHN? This is the Department of Roads. I understand you're looking for work and we have a job opening in our Security Sector. We need someone to find out who keeps putting all those old shoes along the freeway.

* CLICK *

(AS FAST AS POSSIBLE)
Hello. This is the Department of Motor Vehicle Glove Compartments. Our department is calling your apartment to inquire about your car glove compartment. Once a year your car glove compartment has to be brought into our department for inspection. You must comply immediately or our department will come to your apartment and impound your glove compartment. If you have any questions, don't call the Department of Motor Vehicle Glove Compartments, contact the Department of Apartment Owner's Glove Compartment Arguments. Thank you for your cooperation.

∗ CLICK ∗

Hello. This is the Alternative Dental Association. We're calling to introduce you to all-new "Punk-Rock Dentistry":
• No anaesthesia
• We promote and encourage screaming
• The drill bit is dull and the dentist is blind
For an immediate appointment, call 700-LUV-PAIN.

∗ CLICK ∗

Hello Mr. JOHN. This is Warden Smith from the NAME* Correctional Institute. [*PICK THE ONE NEAREST OR MOST WELL KNOWN IN YOUR STATE] I'd like to thank you for volunteering for our new "Adopt-A-Hardened-Criminal" experimental rehabilitation program. Your convict for this weekend is Chainsaw Charlie, a mass murderer. When he is dropped off at your home on Friday afternoon, here are some helpful hints:
- No *knives* at the dinner table.
- When leaving a room he is in, always *back* out the door.
- *Hide* all your power tools.
- When showering, use the "buddy system"
- Try not to upset him, for example, it's best not to wear *striped* pajamas.
- And finally, as for outdoor activities:
Fishing . . . yes!   Hunting . . . NO!

* CLICK *

Hello? Mr. JOHN? This is Direct Response Mail Advertising. We're calling to make sure you received in the mail your free sample of the new chocolate candy made in San Francisco by transvestites? It's called a "He-She" Bar. If it didn't arrive, call 700-BOY-GIRL.

* CLICK *

Hello? Ms. JANE? Have you ever thrown salt over your left shoulder? Do you carry a rabbit's foot? Do you experience panic at the sight of a broken mirror? Then you should consider coming to a meeting of Superstitions Anonymous. At S.A., we believe it is bad luck to be superstitious! For meeting times, call 700-BLAK-CAT.

* CLICK *

[TWO PEOPLE TALKING—ALTERNATE EVERY OTHER SENTENCE]

(FAST — TV ANNOUNCER-TYPE VOICES)
#1: Hello! This is Dr. Looney!
#2: Hi there! This is Dr. Ben!

(TOGETHER)
We're from the new Looney-Ben Asylum!

#1: Check out our Grand-Opening Specials:
#2: Buy one lobotomy-get one free!
#1: Half off on straight jackets!
#2: Plus, all the tranquilizers you can eat!
#1: Bring the whole family!

(TOGETHER)
You've got to be *insane* to pass up these *crazy* bargains!

#2: Call 700-IMA-NUTT

* CLICK *

This is the Flip Side Funeral Parlor, specializing in the interment of people with a shady past. We bury you face-down, so you can see where you're going. Don't wait till it's too late— call now for a free estimate: 700 GO2-HE**.

* CLICK *

This is Air Quality Management. We're calling to inform you that we have finally traced a particularly foul smelling odor to the message on your machine. To avoid a fine, please take the appropriate measures to correct the problem: sell your machine and hit the shower! We can be reached at 700-YU-STINK.

* CLICK *

(SHOUT EACH WORD)

This   is   the   Hearing   Institute!
For   a   free   exam   call
7 0 0 - S P E A K - U P !

* CLICK *

Mr. JOHN? This is the City Library. We're calling about an overdue *Geography* book you've had checked out since Fourth Grade. By now you must be an expert on the subject, so I don't have to tell you what *creek* you're up!

* CLICK *

ABOUT THE AUTHORS

After flunking out of Mortician's School for eating "pizza with everything" during autopsies, CLIFF CARLE and JOHN CARFI became comedy writers. Both have written for BIG NAME comedians (who must remain anonymous because their lawyers said so). In addition, John Carfi is a stand-up comedian who headlines in the top clubs around the country and appears on television. Cliff Carle lives in L.A., John Carfi lives in Pennsylvania. Their minds are elsewhere.

JOHN CARFI

CLIFF CARLE

TITLES BY CCC PUBLICATIONS

RETAIL $4.99

"?"
POSITIVELY PREGNANT
SIGNS YOUR SEX LIFE IS DEAD
WHY MEN DON'T HAVE A CLUE
40 AND HOLDING YOUR OWN
CAN SEX IMPROVE YOUR GOLF?
THE COMPLETE BOOGER BOOK
THINGS YOU CAN DO WITH A USELESS MAN
FLYING FUNNIES
MARITAL BLISS & OXYMORONS
THE VERY VERY SEXY ADULT DOT-TO-DOT BOOK
THE DEFINITIVE FART BOOK
THE COMPLETE WIMP'S GUIDE TO SEX
THE CAT OWNER'S SHAPE UP MANUAL
PMS CRAZED: TOUCH ME AND I'LL KILL YOU!
RETIRED: LET THE GAMES BEGIN
MALE BASHING: WOMEN'S FAVORITE PASTIME
THE OFFICE FROM HELL
FOOD & SEX
FITNESS FANATICS
YOUNGER MEN ARE BETTER THAN RETIN-A
BUT OSSIFER, IT'S NOT MY FAULT

RETAIL $4.95

1001 WAYS TO PROCRASTINATE

THE WORLD'S GREATEST PUT-DOWN LINES

HORMONES FROM HELL II

SHARING THE ROAD WITH IDIOTS

THE GREATEST ANSWERING MACHINE MESSAGES
OF ALL TIME

WHAT DO WE DO NOW?? (A Guide For New Parents)

HOW TO TALK YOU WAY OUT OF A TRAFFIC TICKET

THE BOTTOM HALF (How To Spot Incompetent
Professionals)

LIFE'S MOST EMBARRASSING MOMENTS

HOW TO ENTERTAIN PEOPLE YOU HATE

YOUR GUIDE TO CORPORATE SURVIVAL

THE SUPERIOR PERSON'S GUIDE TO EVERYDAY
IRRITATIONS

GIFTING RIGHT

RETAIL $3.95

YOU KNOW YOU'RE AN OLD FART WHEN...

NO HANG-UPS

NO HANG-UPS II

NO HANG-UPS III

HOW TO SUCCEED IN SINGLES BARS

HOW TO GET EVEN WITH YOUR EXES

TOTALLY OUTRAGEOUS BUMPER-SNICKERS ($2.95)

RETAIL $5.95

LITTLE INSTRUCTION BOOK OF THE RICH & FAMOUS
GETTING EVEN WITH THE ANSWERING MACHINE
ARE YOU A SPORTS NUT?
MEN ARE PIGS / WOMEN ARE BITCHES
50 WAYS TO HUSTLE YOUR FRIENDS ($5.99)
HORMONES FROM HELL
HUSBANDS FROM HELL
KILLER BRAS & Other Hazards Of The 50's
IT'S BETTER TO BE OVER THE HILL THAN UNDER IT
HOW TO REALLY PARTY!!!
WORK SUCKS!
THE PEOPLE WATCHER'S FIRLD GUIDE
THE UNOFFICIAL WOMEN'S DIVORCE GUIDE
THE ABSOLUTE LAST CHANCE DIET BOOK
FOR MEN ONLY (How To Survive Marriage)
THE UGLY TRUTH ABOUT MEN
NEVER A DULL CARD
RED HOT MONOGAMY
   (In Just 60 Seconds A Day) ($6.95)

NO HANG-UPS – CASSETTES Retail $4.98

Vol. I: GENERAL MESSAGES (Female)
Vol. I: GENERAL MESSAGES (Male)
Vol. II: BUSINESS MESSAGES (Female)
Vol. II: BUSINESS MESSAGES (Male)
Vol. III:'R' RATED MESSAGES (Female)
Vol. III:'R' RATED MESSAGES (Male)
Vol. IV:SOUND EFFECTS ONLY
Vol. V: CELEBRI-TEASE